THIS BOOK BELONGS TO:

Copyright © Teresa Rother
All rights reserved. No part of this publication may be reproduced, distributed, or transmitted in any form or by any means, including photocopy, recording, or other electronic or mechanical methods.

DEDICATION

This Chicken Log Book is dedicated to backyard birders, homesteaders, small farms, and businesses who want to manage and track the health of their flock, keep accurate incubation records, and hatching information.

You are my inspiration for producing this book and I'm honored to be a part of helping you manage and retain important information regarding your chickens.

HOW TO USE THIS BOOK

This Chicken Record Book will help you record, collect, and organize your information in an easy to use format.

Here are examples of information for you to fill in and write the details of raising your chickens.

Fill in the following information:

1. Planning Calendar for 12 months

2. Chicken Health Log - record name, breed, egg color health, medication/treatments, vet appointments, a place for notes, and a photo

3. Chicken Hatching Log - 21-day turning schedule- record date started, hatch date, number of egg, a checklist for each day of hatching (egg turning, temperature, and humidity), a reminder on day 18 to stop turning eggs

PLANNING CALENDAR

JANUARY

FEBRUARY

MARCH

APRIL

MAY

JUNE

PLANNING CALENDAR

JULY

AUGUST

SEPTEMBER

OCTOBER

NOVEMBER

DECEMBER

CHICKEN HEALTH LOG

PHOTO

CHICKEN'S NAME	
BREED	
EGG COLOR	
HEALTH	
MEDICATIONS/TREATMENTS	
VET APPOINTMENTS	

NOTES

CHICKEN HEALTH LOG

[PHOTO]

CHICKEN'S NAME	
BREED	
EGG COLOR	
HEALTH	
MEDICATIONS/TREATMENTS	
VET APPOINTMENTS	

NOTES

CHICKEN HEALTH LOG

PHOTO

CHICKEN'S NAME	
BREED	
EGG COLOR	
HEALTH	
MEDICATIONS/TREATMENTS	
VET APPOINTMENTS	

NOTES

CHICKEN HEALTH LOG

PHOTO

CHICKEN'S NAME	
BREED	
EGG COLOR	
HEALTH	
MEDICATIONS/TREATMENTS	
VET APPOINTMENTS	

NOTES

CHICKEN HEALTH LOG

PHOTO

CHICKEN'S NAME	
BREED	
EGG COLOR	
HEALTH	
MEDICATIONS/TREATMENTS	
VET APPOINTMENTS	

NOTES

CHICKEN HEALTH LOG

PHOTO

CHICKEN'S NAME	
BREED	
EGG COLOR	
HEALTH	
MEDICATIONS/TREATMENTS	
VET APPOINTMENTS	

NOTES

CHICKEN HEALTH LOG

PHOTO

CHICKEN'S NAME	
BREED	
EGG COLOR	
HEALTH	
MEDICATIONS/TREATMENTS	
VET APPOINTMENTS	

NOTES

CHICKEN HEALTH LOG

PHOTO

CHICKEN'S NAME	
BREED	
EGG COLOR	
HEALTH	
MEDICATIONS/TREATMENTS	
VET APPOINTMENTS	

NOTES

CHICKEN HEALTH LOG

PHOTO

CHICKEN'S NAME	
BREED	
EGG COLOR	
HEALTH	
MEDICATIONS/TREATMENTS	
VET APPOINTMENTS	

NOTES

CHICKEN HEALTH LOG

PHOTO

CHICKEN'S NAME	
BREED	
EGG COLOR	
HEALTH	
MEDICATIONS/TREATMENTS	
VET APPOINTMENTS	

NOTES

CHICKEN HEALTH LOG

PHOTO

CHICKEN'S NAME	
BREED	
EGG COLOR	
HEALTH	
MEDICATIONS/TREATMENTS	
VET APPOINTMENTS	

NOTES

CHICKEN HEALTH LOG

PHOTO

CHICKEN'S NAME	
BREED	
EGG COLOR	
HEALTH	
MEDICATIONS/TREATMENTS	
VET APPOINTMENTS	

NOTES

CHICKEN HEALTH LOG

PHOTO

CHICKEN'S NAME	
BREED	
EGG COLOR	
HEALTH	
MEDICATIONS/TREATMENTS	
VET APPOINTMENTS	

NOTES

CHICKEN HEALTH LOG

PHOTO

CHICKEN'S NAME	
BREED	
EGG COLOR	
HEALTH	
MEDICATIONS/TREATMENTS	
VET APPOINTMENTS	

NOTES

CHICKEN HEALTH LOG

PHOTO

CHICKEN'S NAME	
BREED	
EGG COLOR	
HEALTH	
MEDICATIONS/TREATMENTS	
VET APPOINTMENTS	

NOTES

CHICKEN HEALTH LOG

PHOTO

CHICKEN'S NAME	
BREED	
EGG COLOR	
HEALTH	
MEDICATIONS/TREATMENTS	
VET APPOINTMENTS	

NOTES

CHICKEN HEALTH LOG

PHOTO

CHICKEN'S NAME	
BREED	
EGG COLOR	
HEALTH	
MEDICATIONS/TREATMENTS	
VET APPOINTMENTS	

NOTES

CHICKEN HEALTH LOG

PHOTO

CHICKEN'S NAME	
BREED	
EGG COLOR	
HEALTH	
MEDICATIONS/TREATMENTS	
VET APPOINTMENTS	

NOTES

CHICKEN HEALTH LOG

[PHOTO]

CHICKEN'S NAME	
BREED	
EGG COLOR	
HEALTH	
MEDICATIONS/TREATMENTS	
VET APPOINTMENTS	

NOTES

CHICKEN HEALTH LOG

PHOTO

CHICKEN'S NAME	
BREED	
EGG COLOR	
HEALTH	
MEDICATIONS/TREATMENTS	
VET APPOINTMENTS	

NOTES

CHICKEN HEALTH LOG

PHOTO

CHICKEN'S NAME	
BREED	
EGG COLOR	
HEALTH	
MEDICATIONS/TREATMENTS	
VET APPOINTMENTS	

NOTES

CHICKEN HEALTH LOG

PHOTO

CHICKEN'S NAME	
BREED	
EGG COLOR	
HEALTH	
MEDICATIONS/TREATMENTS	
VET APPOINTMENTS	

NOTES

CHICKEN HEALTH LOG

PHOTO

CHICKEN'S NAME	
BREED	
EGG COLOR	
HEALTH	
MEDICATIONS/TREATMENTS	
VET APPOINTMENTS	

NOTES

CHICKEN HEALTH LOG

PHOTO

CHICKEN'S NAME	
BREED	
EGG COLOR	
HEALTH	
MEDICATIONS/TREATMENTS	
VET APPOINTMENTS	

NOTES

CHICKEN HEALTH LOG

PHOTO

CHICKEN'S NAME	
BREED	
EGG COLOR	
HEALTH	
MEDICATIONS/TREATMENTS	
VET APPOINTMENTS	

NOTES

CHICKEN HEALTH LOG

PHOTO

CHICKEN'S NAME	
BREED	
EGG COLOR	
HEALTH	
MEDICATIONS/TREATMENTS	
VET APPOINTMENTS	

NOTES

CHICKEN HEALTH LOG

PHOTO

CHICKEN'S NAME	
BREED	
EGG COLOR	
HEALTH	
MEDICATIONS/TREATMENTS	
VET APPOINTMENTS	

NOTES

CHICKEN HEALTH LOG

PHOTO

CHICKEN'S NAME	
BREED	
EGG COLOR	
HEALTH	
MEDICATIONS/TREATMENTS	
VET APPOINTMENTS	

NOTES

CHICKEN HEALTH LOG

PHOTO

CHICKEN'S NAME	
BREED	
EGG COLOR	
HEALTH	
MEDICATIONS/TREATMENTS	
VET APPOINTMENTS	

NOTES

CHICKEN HEALTH LOG

PHOTO

CHICKEN'S NAME	
BREED	
EGG COLOR	
HEALTH	
MEDICATIONS/TREATMENTS	
VET APPOINTMENTS	

NOTES

CHICKEN HEALTH LOG

PHOTO

CHICKEN'S NAME	
BREED	
EGG COLOR	
HEALTH	
MEDICATIONS/TREATMENTS	
VET APPOINTMENTS	

NOTES

CHICKEN HEALTH LOG

PHOTO

CHICKEN'S NAME	
BREED	
EGG COLOR	
HEALTH	
MEDICATIONS/TREATMENTS	
VET APPOINTMENTS	

NOTES

CHICKEN HEALTH LOG

PHOTO

CHICKEN'S NAME	
BREED	
EGG COLOR	
HEALTH	
MEDICATIONS/TREATMENTS	
VET APPOINTMENTS	

NOTES

CHICKEN HEALTH LOG

PHOTO

CHICKEN'S NAME	
BREED	
EGG COLOR	
HEALTH	
MEDICATIONS/TREATMENTS	
VET APPOINTMENTS	

NOTES

CHICKEN HEALTH LOG

PHOTO

CHICKEN'S NAME	
BREED	
EGG COLOR	
HEALTH	
MEDICATIONS/TREATMENTS	
VET APPOINTMENTS	

NOTES

CHICKEN HEALTH LOG

PHOTO

CHICKEN'S NAME	
BREED	
EGG COLOR	
HEALTH	
MEDICATIONS/TREATMENTS	
VET APPOINTMENTS	

NOTES

CHICKEN HEALTH LOG

PHOTO

CHICKEN'S NAME	
BREED	
EGG COLOR	
HEALTH	
MEDICATIONS/TREATMENTS	
VET APPOINTMENTS	

NOTES

CHICKEN HEALTH LOG

PHOTO

CHICKEN'S NAME	
BREED	
EGG COLOR	
HEALTH	
MEDICATIONS/TREATMENTS	
VET APPOINTMENTS	

NOTES

CHICKEN HEALTH LOG

PHOTO

CHICKEN'S NAME	
BREED	
EGG COLOR	
HEALTH	
MEDICATIONS/TREATMENTS	
VET APPOINTMENTS	

NOTES

CHICKEN HEALTH LOG

PHOTO

CHICKEN'S NAME	
BREED	
EGG COLOR	
HEALTH	
MEDICATIONS/TREATMENTS	
VET APPOINTMENTS	

NOTES

CHICKEN HEALTH LOG

PHOTO

CHICKEN'S NAME	
BREED	
EGG COLOR	
HEALTH	
MEDICATIONS/TREATMENTS	
VET APPOINTMENTS	

NOTES

CHICKEN HEALTH LOG

PHOTO

CHICKEN'S NAME	
BREED	
EGG COLOR	
HEALTH	
MEDICATIONS/TREATMENTS	
VET APPOINTMENTS	

NOTES

CHICKEN HEALTH LOG

PHOTO

CHICKEN'S NAME	
BREED	
EGG COLOR	
HEALTH	
MEDICATIONS/TREATMENTS	
VET APPOINTMENTS	

NOTES

CHICKEN HEALTH LOG

PHOTO

CHICKEN'S NAME	
BREED	
EGG COLOR	
HEALTH	
MEDICATIONS/TREATMENTS	
VET APPOINTMENTS	

NOTES

CHICKEN HEALTH LOG

PHOTO

CHICKEN'S NAME	
BREED	
EGG COLOR	
HEALTH	
MEDICATIONS/TREATMENTS	
VET APPOINTMENTS	

NOTES

CHICKEN HEALTH LOG

PHOTO

CHICKEN'S NAME	
BREED	
EGG COLOR	
HEALTH	
MEDICATIONS/TREATMENTS	
VET APPOINTMENTS	

NOTES

CHICKEN HEALTH LOG

PHOTO

CHICKEN'S NAME	
BREED	
EGG COLOR	
HEALTH	
MEDICATIONS/TREATMENTS	
VET APPOINTMENTS	

NOTES

CHICKEN HEALTH LOG

PHOTO

CHICKEN'S NAME	
BREED	
EGG COLOR	
HEALTH	
MEDICATIONS/TREATMENTS	
VET APPOINTMENTS	

NOTES

CHICKEN HEALTH LOG

PHOTO

CHICKEN'S NAME	
BREED	
EGG COLOR	
HEALTH	
MEDICATIONS/TREATMENTS	
VET APPOINTMENTS	

NOTES

CHICKEN HEALTH LOG

[PHOTO]

CHICKEN'S NAME	
BREED	
EGG COLOR	
HEALTH	
MEDICATIONS/TREATMENTS	
VET APPOINTMENTS	

NOTES

CHICKEN HATCHING LOG

START DATE _____ HATCH DATE _____ NUMBER OF EGGS _____

DAY	TURN	CHECK TEMP/HUM	TURN	TURN	TURN
1					
2					
3					
4					
5					
6					
7					
8					
9					
10					
11					
12					
13					
14					
15					
16					
17					
18- STOP TURNING EGGS	———	———	———	———	———
19					
20					
21					

CHICKEN HATCHING LOG

START DATE _____ HATCH DATE _____ NUMBER OF EGGS _____

DAY	TURN	CHECK TEMP/HUM	TURN	TURN	TURN
1					
2					
3					
4					
5					
6					
7					
8					
9					
10					
11					
12					
13					
14					
15					
16					
17					
18- STOP TURNING EGGS	_____	_____	_____	_____	_____
19					
20					
21					

CHICKEN HATCHING LOG

START DATE _____ HATCH DATE _____ NUMBER OF EGGS _____

DAY	TURN	CHECK TEMP/HUM	TURN	TURN	TURN
1					
2					
3					
4					
5					
6					
7					
8					
9					
10					
11					
12					
13					
14					
15					
16					
17					
18- STOP TURNING EGGS	———	———	———	———	———
19					
20					
21					

CHICKEN HATCHING LOG

START DATE _____ HATCH DATE _____ NUMBER OF EGGS _____

DAY	TURN	CHECK TEMP/HUM	TURN	TURN	TURN
1					
2					
3					
4					
5					
6					
7					
8					
9					
10					
11					
12					
13					
14					
15					
16					
17					
18- STOP TURNING EGGS	_____	_____	_____	_____	_____
19					
20					
21					

CHICKEN HATCHING LOG

START DATE _____ HATCH DATE _____ NUMBER OF EGGS _____

DAY	TURN	CHECK TEMP/HUM	TURN	TURN	TURN
1					
2					
3					
4					
5					
6					
7					
8					
9					
10					
11					
12					
13					
14					
15					
16					
17					
18- STOP TURNING EGGS	___	___	___	___	___
19					
20					
21					

CHICKEN HATCHING LOG

START DATE _____ HATCH DATE _____ NUMBER OF EGGS _____

DAY	TURN	CHECK TEMP/HUM	TURN	TURN	TURN
1					
2					
3					
4					
5					
6					
7					
8					
9					
10					
11					
12					
13					
14					
15					
16					
17					
18- STOP TURNING EGGS	_____	_____	_____	_____	_____
19					
20					
21					

CHICKEN HATCHING LOG

START DATE _____ HATCH DATE _____ NUMBER OF EGGS _____

DAY	TURN	CHECK TEMP/HUM	TURN	TURN	TURN
1					
2					
3					
4					
5					
6					
7					
8					
9					
10					
11					
12					
13					
14					
15					
16					
17					
18- STOP TURNING EGGS	―――	―――	―――	―――	―――
19					
20					
21					

CHICKEN HATCHING LOG

START DATE _____ HATCH DATE _____ NUMBER OF EGGS _____

DAY	TURN	CHECK TEMP/HUM	TURN	TURN	TURN
1					
2					
3					
4					
5					
6					
7					
8					
9					
10					
11					
12					
13					
14					
15					
16					
17					
18- STOP TURNING EGGS	_____	_____	_____	_____	_____
19					
20					
21					

CHICKEN HATCHING LOG

START DATE _____ HATCH DATE _____ NUMBER OF EGGS _____

DAY	TURN	CHECK TEMP/HUM	TURN	TURN	TURN
1					
2					
3					
4					
5					
6					
7					
8					
9					
10					
11					
12					
13					
14					
15					
16					
17					
18- STOP TURNING EGGS					
19					
20					
21					

CHICKEN HATCHING LOG

START DATE _____ HATCH DATE _____ NUMBER OF EGGS _____

DAY	TURN	CHECK TEMP/HUM	TURN	TURN	TURN
1					
2					
3					
4					
5					
6					
7					
8					
9					
10					
11					
12					
13					
14					
15					
16					
17					
18- STOP TURNING EGGS	___	___	___	___	___
19					
20					
21					

CHICKEN HATCHING LOG

START DATE _____ HATCH DATE _____ NUMBER OF EGGS _____

DAY	TURN	CHECK TEMP/HUM	TURN	TURN	TURN
1					
2					
3					
4					
5					
6					
7					
8					
9					
10					
11					
12					
13					
14					
15					
16					
17					
18- STOP TURNING EGGS	-----	-----	-----	-----	-----
19					
20					
21					

CHICKEN HATCHING LOG

START DATE _____ HATCH DATE _____ NUMBER OF EGGS _____

DAY	TURN	CHECK TEMP/HUM	TURN	TURN	TURN
1					
2					
3					
4					
5					
6					
7					
8					
9					
10					
11					
12					
13					
14					
15					
16					
17					
18- STOP TURNING EGGS	___	___	___	___	___
19					
20					
21					

CHICKEN HATCHING LOG

START DATE _____ HATCH DATE _____ NUMBER OF EGGS _____

DAY	TURN	CHECK TEMP/HUM	TURN	TURN	TURN
1					
2					
3					
4					
5					
6					
7					
8					
9					
10					
11					
12					
13					
14					
15					
16					
17					
18- STOP TURNING EGGS	———	———	———	———	———
19					
20					
21					

CHICKEN HATCHING LOG

START DATE _____ HATCH DATE _____ NUMBER OF EGGS _____

DAY	TURN	CHECK TEMP/HUM	TURN	TURN	TURN
1					
2					
3					
4					
5					
6					
7					
8					
9					
10					
11					
12					
13					
14					
15					
16					
17					
18- STOP TURNING EGGS	_____	_____	_____	_____	_____
19					
20					
21					

CHICKEN HATCHING LOG

START DATE _____ HATCH DATE _____ NUMBER OF EGGS _____

DAY	TURN	CHECK TEMP/HUM	TURN	TURN	TURN
1					
2					
3					
4					
5					
6					
7					
8					
9					
10					
11					
12					
13					
14					
15					
16					
17					
18- STOP TURNING EGGS	————	————	————	————	————
19					
20					
21					

CHICKEN HATCHING LOG

START DATE _____ HATCH DATE _____ NUMBER OF EGGS _____

DAY	TURN	CHECK TEMP/HUM	TURN	TURN	TURN
1					
2					
3					
4					
5					
6					
7					
8					
9					
10					
11					
12					
13					
14					
15					
16					
17					
18- STOP TURNING EGGS	_____	_____	_____	_____	_____
19					
20					
21					

CHICKEN HATCHING LOG

START DATE _____ HATCH DATE _____ NUMBER OF EGGS _____

DAY	TURN	CHECK TEMP/HUM	TURN	TURN	TURN
1					
2					
3					
4					
5					
6					
7					
8					
9					
10					
11					
12					
13					
14					
15					
16					
17					
18- STOP TURNING EGGS	------	------	------	------	------
19					
20					
21					

CHICKEN HATCHING LOG

START DATE _____ HATCH DATE _____ NUMBER OF EGGS _____

DAY	TURN	CHECK TEMP/HUM	TURN	TURN	TURN
1					
2					
3					
4					
5					
6					
7					
8					
9					
10					
11					
12					
13					
14					
15					
16					
17					
18- STOP TURNING EGGS	------	------	------	------	------
19					
20					
21					

CHICKEN HATCHING LOG

START DATE _____ HATCH DATE _____ NUMBER OF EGGS _____

DAY	TURN	CHECK TEMP/HUM	TURN	TURN	TURN
1					
2					
3					
4					
5					
6					
7					
8					
9					
10					
11					
12					
13					
14					
15					
16					
17					
18- STOP TURNING EGGS					
19					
20					
21					

CHICKEN HATCHING LOG

START DATE _____ HATCH DATE _____ NUMBER OF EGGS _____

DAY	TURN	CHECK TEMP/HUM	TURN	TURN	TURN
1					
2					
3					
4					
5					
6					
7					
8					
9					
10					
11					
12					
13					
14					
15					
16					
17					
18- STOP TURNING EGGS	---	---	---	---	---
19					
20					
21					

CHICKEN HATCHING LOG

START DATE _____ HATCH DATE _____ NUMBER OF EGGS _____

DAY	TURN	CHECK TEMP/HUM	TURN	TURN	TURN
1					
2					
3					
4					
5					
6					
7					
8					
9					
10					
11					
12					
13					
14					
15					
16					
17					
18- STOP TURNING EGGS	—	—	—	—	—
19					
20					
21					

CHICKEN HATCHING LOG

START DATE _____ HATCH DATE _____ NUMBER OF EGGS _____

DAY	TURN	CHECK TEMP/HUM	TURN	TURN	TURN
1					
2					
3					
4					
5					
6					
7					
8					
9					
10					
11					
12					
13					
14					
15					
16					
17					
18 - STOP TURNING EGGS	———	———	———	———	———
19					
20					
21					

CHICKEN HATCHING LOG

START DATE _____ HATCH DATE _____ NUMBER OF EGGS _____

DAY	TURN	CHECK TEMP/HUM	TURN	TURN	TURN
1					
2					
3					
4					
5					
6					
7					
8					
9					
10					
11					
12					
13					
14					
15					
16					
17					
18- STOP TURNING EGGS					
19					
20					
21					

CHICKEN HATCHING LOG

START DATE _____ HATCH DATE _____ NUMBER OF EGGS _____

DAY	TURN	CHECK TEMP/HUM	TURN	TURN	TURN
1					
2					
3					
4					
5					
6					
7					
8					
9					
10					
11					
12					
13					
14					
15					
16					
17					
18- STOP TURNING EGGS	———	———	———	———	———
19					
20					
21					

CHICKEN HATCHING LOG

START DATE _____ HATCH DATE _____ NUMBER OF EGGS _____

DAY	TURN	CHECK TEMP/HUM	TURN	TURN	TURN
1					
2					
3					
4					
5					
6					
7					
8					
9					
10					
11					
12					
13					
14					
15					
16					
17					
18- STOP TURNING EGGS					
19					
20					
21					

CHICKEN HATCHING LOG

START DATE _____ HATCH DATE _____ NUMBER OF EGGS _____

DAY	TURN	CHECK TEMP/HUM	TURN	TURN	TURN
1					
2					
3					
4					
5					
6					
7					
8					
9					
10					
11					
12					
13					
14					
15					
16					
17					
18- STOP TURNING EGGS	_____	_____	_____	_____	_____
19					
20					
21					

CHICKEN HATCHING LOG

START DATE _____ HATCH DATE _____ NUMBER OF EGGS _____

DAY	TURN	CHECK TEMP/HUM	TURN	TURN	TURN
1					
2					
3					
4					
5					
6					
7					
8					
9					
10					
11					
12					
13					
14					
15					
16					
17					
18- STOP TURNING EGGS					
19					
20					
21					

CHICKEN HATCHING LOG

START DATE _____ HATCH DATE _____ NUMBER OF EGGS _____

DAY	TURN	CHECK TEMP/HUM	TURN	TURN	TURN
1					
2					
3					
4					
5					
6					
7					
8					
9					
10					
11					
12					
13					
14					
15					
16					
17					
18- STOP TURNING EGGS	------	------	------	------	------
19					
20					
21					

CHICKEN HATCHING LOG

START DATE _____ HATCH DATE _____ NUMBER OF EGGS _____

DAY	TURN	CHECK TEMP/HUM	TURN	TURN	TURN
1					
2					
3					
4					
5					
6					
7					
8					
9					
10					
11					
12					
13					
14					
15					
16					
17					
18- STOP TURNING EGGS	_____	_____	_____	_____	_____
19					
20					
21					

CHICKEN HATCHING LOG

START DATE _____ HATCH DATE _____ NUMBER OF EGGS _____

DAY	TURN	CHECK TEMP/HUM	TURN	TURN	TURN
1					
2					
3					
4					
5					
6					
7					
8					
9					
10					
11					
12					
13					
14					
15					
16					
17					
18- STOP TURNING EGGS	___	___	___	___	___
19					
20					
21					

CHICKEN HATCHING LOG

START DATE _____ HATCH DATE _____ NUMBER OF EGGS _____

DAY	TURN	CHECK TEMP/HUM	TURN	TURN	TURN
1					
2					
3					
4					
5					
6					
7					
8					
9					
10					
11					
12					
13					
14					
15					
16					
17					
18- STOP TURNING EGGS	———	———	———	———	———
19					
20					
21					

CHICKEN HATCHING LOG

START DATE _____ HATCH DATE _____ NUMBER OF EGGS _____

DAY	TURN	CHECK TEMP/HUM	TURN	TURN	TURN
1					
2					
3					
4					
5					
6					
7					
8					
9					
10					
11					
12					
13					
14					
15					
16					
17					
18- STOP TURNING EGGS	-------	-------	-------	-------	-------
19					
20					
21					

CHICKEN HATCHING LOG

START DATE _____ HATCH DATE _____ NUMBER OF EGGS _____

DAY	TURN	CHECK TEMP/HUM	TURN	TURN	TURN
1					
2					
3					
4					
5					
6					
7					
8					
9					
10					
11					
12					
13					
14					
15					
16					
17					
18- STOP TURNING EGGS	_____	_____	_____	_____	_____
19					
20					
21					

CHICKEN HATCHING LOG

START DATE _____ HATCH DATE _____ NUMBER OF EGGS _____

DAY	TURN	CHECK TEMP/HUM	TURN	TURN	TURN
1					
2					
3					
4					
5					
6					
7					
8					
9					
10					
11					
12					
13					
14					
15					
16					
17					
18- STOP TURNING EGGS	_____	_____	_____	_____	_____
19					
20					
21					

CHICKEN HATCHING LOG

START DATE _____ HATCH DATE _____ NUMBER OF EGGS _____

DAY	TURN	CHECK TEMP/HUM	TURN	TURN	TURN
1					
2					
3					
4					
5					
6					
7					
8					
9					
10					
11					
12					
13					
14					
15					
16					
17					
18- STOP TURNING EGGS					
19					
20					
21					

CHICKEN HATCHING LOG

START DATE _____ HATCH DATE _____ NUMBER OF EGGS _____

DAY	TURN	CHECK TEMP/HUM	TURN	TURN	TURN
1					
2					
3					
4					
5					
6					
7					
8					
9					
10					
11					
12					
13					
14					
15					
16					
17					
18- STOP TURNING EGGS	_____	_____	_____	_____	_____
19					
20					
21					

CHICKEN HATCHING LOG

START DATE _____ HATCH DATE _____ NUMBER OF EGGS _____

DAY	TURN	CHECK TEMP/HUM	TURN	TURN	TURN
1					
2					
3					
4					
5					
6					
7					
8					
9					
10					
11					
12					
13					
14					
15					
16					
17					
18- STOP TURNING EGGS					
19					
20					
21					

CHICKEN HATCHING LOG

START DATE _____ HATCH DATE _____ NUMBER OF EGGS _____

DAY	TURN	CHECK TEMP/HUM	TURN	TURN	TURN
1					
2					
3					
4					
5					
6					
7					
8					
9					
10					
11					
12					
13					
14					
15					
16					
17					
18- STOP TURNING EGGS	___	___	___	___	___
19					
20					
21					

CHICKEN HATCHING LOG

START DATE _____ HATCH DATE _____ NUMBER OF EGGS _____

DAY	TURN	CHECK TEMP/HUM	TURN	TURN	TURN
1					
2					
3					
4					
5					
6					
7					
8					
9					
10					
11					
12					
13					
14					
15					
16					
17					
18- STOP TURNING EGGS	_____	_____		_____	_____
19					
20					
21					

CHICKEN HATCHING LOG

START DATE _____ HATCH DATE _____ NUMBER OF EGGS _____

DAY	TURN	CHECK TEMP/HUM	TURN	TURN	TURN
1					
2					
3					
4					
5					
6					
7					
8					
9					
10					
11					
12					
13					
14					
15					
16					
17					
18- STOP TURNING EGGS	_____	_____	_____	_____	_____
19					
20					
21					

CHICKEN HATCHING LOG

START DATE _____ HATCH DATE _____ NUMBER OF EGGS _____

DAY	TURN	CHECK TEMP/HUM	TURN	TURN	TURN
1					
2					
3					
4					
5					
6					
7					
8					
9					
10					
11					
12					
13					
14					
15					
16					
17					
18- STOP TURNING EGGS	―――	―――	―――	―――	―――
19					
20					
21					

CHICKEN HATCHING LOG

START DATE _____ HATCH DATE _____ NUMBER OF EGGS _____

DAY	TURN	CHECK TEMP/HUM	TURN	TURN	TURN
1					
2					
3					
4					
5					
6					
7					
8					
9					
10					
11					
12					
13					
14					
15					
16					
17					
18- STOP TURNING EGGS	_____	_____	_____	_____	_____
19					
20					
21					

CHICKEN HATCHING LOG

START DATE _____ HATCH DATE _____ NUMBER OF EGGS _____

DAY	TURN	CHECK TEMP/HUM	TURN	TURN	TURN
1					
2					
3					
4					
5					
6					
7					
8					
9					
10					
11					
12					
13					
14					
15					
16					
17					
18- STOP TURNING EGGS	_____	_____	_____	_____	_____
19					
20					
21					

CHICKEN HATCHING LOG

START DATE _____ HATCH DATE _____ NUMBER OF EGGS _____

DAY	TURN	CHECK TEMP/HUM	TURN	TURN	TURN
1					
2					
3					
4					
5					
6					
7					
8					
9					
10					
11					
12					
13					
14					
15					
16					
17					
18- STOP TURNING EGGS	_____	_____	_____	_____	_____
19					
20					
21					

CHICKEN HATCHING LOG

START DATE _____ HATCH DATE _____ NUMBER OF EGGS _____

DAY	TURN	CHECK TEMP/HUM	TURN	TURN	TURN
1					
2					
3					
4					
5					
6					
7					
8					
9					
10					
11					
12					
13					
14					
15					
16					
17					
18- STOP TURNING EGGS	_____	_____	_____	_____	_____
19					
20					
21					

CHICKEN HATCHING LOG

START DATE _____ HATCH DATE _____ NUMBER OF EGGS _____

DAY	TURN	CHECK TEMP/HUM	TURN	TURN	TURN
1					
2					
3					
4					
5					
6					
7					
8					
9					
10					
11					
12					
13					
14					
15					
16					
17					
18- STOP TURNING EGGS	_____	_____	_____	_____	_____
19					
20					
21					

CHICKEN HATCHING LOG

START DATE _____ HATCH DATE _____ NUMBER OF EGGS _____

DAY	TURN	CHECK TEMP/HUM	TURN	TURN	TURN
1					
2					
3					
4					
5					
6					
7					
8					
9					
10					
11					
12					
13					
14					
15					
16					
17					
18- STOP TURNING EGGS					
19					
20					
21					

CHICKEN HATCHING LOG

START DATE _____ HATCH DATE _____ NUMBER OF EGGS _____

DAY	TURN	CHECK TEMP/HUM	TURN	TURN	TURN
1					
2					
3					
4					
5					
6					
7					
8					
9					
10					
11					
12					
13					
14					
15					
16					
17					
18- STOP TURNING EGGS	_____	_____	_____	_____	_____
19					
20					
21					

CHICKEN HATCHING LOG

START DATE _____ HATCH DATE _____ NUMBER OF EGGS _____

DAY	TURN	CHECK TEMP/HUM	TURN	TURN	TURN
1					
2					
3					
4					
5					
6					
7					
8					
9					
10					
11					
12					
13					
14					
15					
16					
17					
18- STOP TURNING EGGS	------	------	------	------	------
19					
20					
21					

CHICKEN HATCHING LOG

START DATE _____ HATCH DATE _____ NUMBER OF EGGS _____

DAY	TURN	CHECK TEMP/HUM	TURN	TURN	TURN
1					
2					
3					
4					
5					
6					
7					
8					
9					
10					
11					
12					
13					
14					
15					
16					
17					
18- STOP TURNING EGGS	_____	_____	_____	_____	_____
19					
20					
21					

CPSIA information can be obtained
at www.ICGtesting.com
Printed in the USA
LVHW062101310121
677963LV00004B/58